YOUR KNOWLEDGE HAS VALUE

- We will publish your bachelor's and master's thesis, essays and papers

- Your own eBook and book - sold worldwide in all relevant shops

- Earn money with each sale

Upload your text at www.GRIN.com and publish for free

Benjamin Toric

Can agriculture be sustainable in its present industrial and high tech form or must it "return to the past"?

GRIN Verlag

Bibliografische Information der Deutschen Nationalbibliothek:

Die Deutsche Bibliothek verzeichnet diese Publikation in der Deutschen National-
bibliografie; detaillierte bibliografische Daten sind im Internet über http://dnb.d-
nb.de/ abrufbar.

Imprint:

Copyright © 2007 GRIN Verlag GmbH
Druck und Bindung: Books on Demand GmbH, Norderstedt Germany
ISBN: 978-3-640-38678-9

This book at GRIN:

http://www.grin.com/en/e-book/111064/can-agriculture-be-sustainable-in-its-present-
industrial-and-high-tech

GRIN - Your knowledge has value

Der GRIN Verlag publiziert seit 1998 wissenschaftliche Arbeiten von Studenten, Hochschullehrern und anderen Akademikern als eBook und gedrucktes Buch. Die Verlagswebsite www.grin.com ist die ideale Plattform zur Veröffentlichung von Hausarbeiten, Abschlussarbeiten, wissenschaftlichen Aufsätzen, Dissertationen und Fachbüchern.

Visit us on the internet:

http://www.grin.com/

http://www.facebook.com/grincom

http://www.twitter.com/grin_com

Can agriculture be sustainable in its present industrial and high tech form or must it "return to the past"?

Prepared by Ben Toric

University of Newcastle

1 ABSTRACT

There are many environmental and social costs associated with conventional agriculture or what may be termed as industrial or high tech agriculture. There is now general agreement that conventional agriculture is not sustainable. The paper considers the way agriculture can produce enough food for the rising global population and at the same time maintain environmental quality. Effects of agriculture on the environment and issues which affected agriculture are considered. The paper concludes that sustainable agriculture probably relies on a combination of farming systems - both organic and a more sustainable form of conventional agriculture relying on green technologies. Certification systems play an important role in achieving sustainability objectives as do ecological taxes. The issue of trade liberalisation and North-South relations is likely to remain much more contentious and harder to resolve in the future, particularly equitable food distribution.

2 Introduction

There is now general agreement that conventional agriculture, or what may be termed as industrial or high tech agriculture, is not sustainable (Ho and Ulanowicz, 2005; Soule and Piper, 1992). If conventional agriculture is not sustainable, then how will enough food be produced for the rising global population, which is expected to reach 8.27 million by 2030, according o Bruinsma (2003)? It is through the use of technologies, particularly pesticides and fertilisers, the we have been able to increase land productivity and per-capita food availability despite a reduction in per-capita agricultural land (Smith *et al.* 2007). So far, food production has kept pace with demand (Eickhout *et al.*, 2006). This was achieved at great environmental and social cost, however.

Environmental and social costs associated with conventional agriculture include loss of soil resources (erosion, waterlogging, salinisation), surface and groundwater contamination due to the application of fertilisers and pesticides, pest resistance, lower biodiversity, decline of family farms, poor working conditions of farm labourers, increasing costs of production, and the disintegration of economic and social conditions in rural communities (UCLA, 1997; Ruttan, 1994). With rising global population and therefore higher demand for food, meat products in particular, pressure on the environment will continue to increase.

The United Nations' Millennium Development Goal (MDG) seeks to reduce the number of people suffering from hunger as well as ensure environmental sustainability (Toledo and Burlingame, 2006). The question is what form of agriculture can produce enough food for the rising global population and at the same time maintain environmental quality? Is the answer a "return to the past" or is there a new way forward?

3   Issues in agriculture

Increasing food production to meet the demand of current and future population will be difficult as there are many issues in agriculture, both in terms of agricultural impacts on the environment and the impact of the environment on agriculture.

3.1   Effects of agriculture on the environment

One of the key issues to consider is the availability of arable land and the way this land is used for agriculture. Increased land use allocation conflicts are expected in the future. Loss of agricultural land to urban land uses is a concern in some states, for example in California according to UCLA (1997) as well as in Bosnia and Herzegovina where the author resides. There is also a loss of agricultural land due to soil erosion and desertification. With rising global population, there will also be a further conversion of natural habitat into agricultural land, primarily in developing countries (Hole *et al.* 2005). Due to increasing meat consumption, most of the arable land expansion will be used for the production of animal feedstuffs (Eickhout *et al.*, 2006; Smith *et al.* 2007). Much of this new arable land will come from the clearing of rainforests which conflicts with environmental goals of the MDG (Eickhout *et al.*, 2006; Smith *et al.* 2007). Land use pressures will be particularly high in India and China (Eickhout *et al.*, 2006). Accompanying these land use changes, will probably be an increase in the use of water, fertilizers and pesticides (Hole *et al.* 2005). This in turn has other consequences that are well documented in the literature, and will only be briefly discussed here.

- **Water use**

There are many issues associated with increased use of water for irrigating crops, including: waterlogging, salinisation, reduction in riparian habitat and degradation of water quality (Wichelns and Oster, 2006; Soule and Piper, 1992). A study by Wichelns and Oster (2006) showed that environmental costs of irrigation in parts of California and Arizona are now even exceeding social and economic benefits. Yet, both Wichelns and Oster (2006) as well as others, such as Mwakalila (2006), suggest that the world cannot meet the future demand for food without irrigation. Irrigation, if managed properly, can increase farmer's productivity and income, thus alleviating rural poverty (Mwakalila, 2006). The challenge is to use water in a much more efficient manner to ensure sustainability of the resource as well avoidance of environmental problems that often accompany regions with significant areas under irrigation.

- **Fertilizer use**

Similarly, there is a challenge to provide nutrients to plants, as currently, for example, only about 50% of nitrogen (N) applied on land is utilised by the crops – the rest pollutes the soil, water or air (Eickhout *et al.*, 2006; Soule and Piper, 1992). N use efficiency must improve (Eickhout *et al.*, 2006). The need for improved efficiency also applies to the other macro-nutrient applied on land – phosphorous – which is currently mined unsustainably from phosphate rock (Birch, 1993). Excessive use of N and P may result in eutrophication of rivers and groundwater pollution.

- **Pesticide use**

Following this, there are great concerns if there is a continued increase in the use of pesticides as, for example, it has already been happening in Central America according to Wesseling, *et al.* (2005). Pesticide use is particularly a problem in developing countries as farmers often use more toxic pesticides due to poor pesticide registration procedures (Wesseling, *et al.*, 2005) and pesticides are used and disposed of inappropriately (Dalvie *et al.*, 2006; Soule and Piper, 1992). Even in countries in transition, in the heart of Europe, such as Bosnia and Herzegovina, there are more than a dozen pesticides in use that are banned in the European Union (EU). Furthermore, farmers usually apply them inappropriately, without the use of safety clothing, and ignoring the withholding periods specified on the label. In addition to the above-mentioned problems, pesticide kill non-target species (reducing biodiversity) and cause pests to develop resistance over time (Soule and Piper, 1992).

The challenge remains to keep pests under control and one way forward is though the use of Integrated Pest Management (IPM), which decreases reliance on chemical pest control methods.

It is important to note that the consequences of increased use of water, fertilizers and pesticides are often magnified as each can have a cumulative, add-on, effect on the environment. For example, Mozumber and Berrens (2007) have found that higher fertilizer use increase biodiversity risk. However, biodiversity can also affected by the use of pesticides, which affect non-target species; and irrigation, which can reduce environmental flows in rivers and increase salinity thus affecting river and riparian flora and fauna. Similarly, use of herbicides can reduce vegetative cover thus increasing water runoff and soil erosion which has further downstream ecosystem effects due to altered water quantity and quality.

3.2 Global issues affecting agriculture

It is also important not to forget that agriculture is readily affected by global issues and trends in climate change; decline of fossil fuels; and trade liberalisation and north-south relations.

- **Climate change; and decline of fossil fuels**

There is now strong scientific evidence that climate change is occurring, the outcome possible being catastrophic as it can increase the risk of abrupt and non-linear changes in ecosystems, which could affect their function, biodiversity and productivity (IPPC, 2001). The impact on agriculture is likely to be large, requiring significant adjustments in farming systems to cope with the changing climate. Agricultural land uses also have an effect on the greenhouse effect, accounting for about 14% of total anthropogenic emissions of greenhouse gases (Bouwman, 2001) and therefore agriculture itself has to meet the challenge of increasing current production and becoming less reliant on limited fossil fuels (Soule and Piper, 1992).

- **Trade liberalisation and North-South relations**

According to Ho and Ulanowicz (2005) and Shah (2006), trade liberalization is resulting in over-exploitation of the poor, especially in developing countries, with resultant over-exploitation of natural resources. Rather pessimistically, Ho and Ulanowicz (2005) predict that natural resources will be further exploited which in the end will destroy both the global economy and the earth's ecosystem. The issues that need to be resolved are how to develop trade between the North and the South without increasing pressure on natural resource in developing countries and, at the same time, promote a more equitable distribution of food for a growing world population.

It is within this context of issues that we need to consider avenues for a sustainable agriculture.

4 Way forward

Considering the need to produce more food for a growing population, and environmental problems associated with conventional agriculture, what are the options for a sustainable agriculture? Some, such as Ruttan (1994), argued in 1994 that we were still not ready to embark on implementing a package of sustainable agriculture practices as the concept still needed further research. Ruttan (1994) justifies this pessimism by stating that the IPM concept emerged in the 1960s but it took two decades before there was a package of IPM practices that could be implemented on the farm. Without a doubt, we have improved our knowledge base since 1994, but Ruttan's statement still stands – we indeed do need to conduct a lot more research on this topic. Research should focus on addressing environmental issues associated with agricultural production, improving animal welfare and identifying new agricultural methods and products (Borch, 2007). Smith *et al.* (2007) then calls for a global exchange of information on green technologies for efficient use of land resources and agricultural inputs.

Will all this research in the end suggest that we must "return to the past", for that is the most appropriate solution to sustainable agriculture? If the "return to the past" means abandonment of all modern agricultural technologies, including those employed in organic agriculture, the answer would be a clear no. Traditional agriculture would not be sufficient to meet the needs of increasing population (Shi, 2002). If the "return to the past" means the use of scientific techniques as applied today, combined with indigenous/traditional knowledge (Birch, 1993), then the answer would be partly yes. Indeed, organic farming is the only truly sustainable form of agriculture (Tait and Morris, 2000). But it would too narrow a view to rely only on organic farming as this view does not consider competing interests of land use, labour and the need to feed a growing population (Tait and Morris, 2000). The answer to sustainable agriculture probably relies on a combination of farming systems - both organic and a more sustainable form of conventional agriculture relying on green technologies.

Without improved technologies resulting in increased yields, more natural habitat will need to be converted to agriculture (Ehrlich, 1994).

According to Borch (2007), future farms will most likely become professionally run, science-based competitive enterprises, driven by knowledge and linked to the world markets. Green technologies will be employed in both organic and conventional systems, although differently as it needs to reflect a different set of values attached to each system (Borch, 2007).

What are some of these green technologies? Some of the technologies proposed by Borch (2007) would probably easily adapt in two farming systems: manure technology and biomass technology. These technologies would allow for the recycling of organic matter, thus eliminating the use of inorganic fertilisers and improving soil health. Borch (2007) also offers some other technologies such as plant gene technology, which is currently a lot more contentious. However, others (Mozumber and Berrens, 2007; Ehrlich, 1994) also suggest that biotechnology can contribute to the sustainability of agriculture.

In general, sustainable approaches are those that are the least toxic and least energy intensive, and yet maintain productivity and profitability (UCLA, 1997) and may include the following technologies/techniques:

- Increased reliance on organic fertilizers, manure and mulch to maintain soil health (Shi 2002).
- Integrated Pest Management (Ehrlich, 1994; Wesseling, *et al.*, 2005; Dalvie *et al.*, 2006).
- Crop rotation and effective soil conservation techniques such as reduced tillage (Ehrlich, 1994; Borch, 2007; UCLA, 1997).
- More efficient irrigation (Birch, 1993).
- Improved exchange of information through information and communication technology (ICT) (Borch, 2007).
- Precision farming (Borch, 2007).
- Reduction in post-harvest losses (Ehrlich, 1994).

- Sympathetic management of non-crop habitats and field margins ((Hole *et al.* 2005).

- Mixed/diversified farming (Hole *et al.* 2005; UCLA, 1997).

- Maximum reliance on natural, renewable, and on-farm inputs (UCLA, 1997).

- Minimum reliance on non-renewable energy sources and a substitution of renewable sources or labour to the extent that is economically feasible (UCLA, 1997).

- Recycling human waste (safely) (Eickhout *et al.*, 2006).

- Integrating biodiversity into food security and anti-hunger policies (Toledo and Burlingame, 2006).

- Ecological taxes that will encourage resource efficiency and more realistic pricing on the market (meat should also become more expensive).

Through the implementation of above strategies there can be visible improvements in environmental outcomes. For example, the review of 76 studies show that species biodiversity, at every level of the food chain, tends to be higher on organic farms that conventional farms (Hole *et al.* 2005). However, increased biodiversity is possible even on conventional farms if appropriate changes in farming practices are implemented, such as substitution of inorganic fertilisers with organic fertilisers and implementation of IPM practices (Hole *et al.* 2005).

Many of the above techniques are strongly encouraged or even mandatory according to certain production standards such as organic or EUREPGAP standards of production. Consumers are increasingly interested in environmental issues and are demanding certification of products to ensure it meets safety, environmental and social criteria; although a study in the Netherlands suggests that most consumers are not well informed about the differences between conventional and organic farming (Hoogland *et al.*, 2007). The benefit of certification is that it encourages a change in production and management practices (Kilian *et al.* 2006), and EUREPGAP standard for example requires the use of IPM and pesticides which are only permitted in the EU. Introduction of EUREPGAP in Bosnia and Herzegovina, for example, has resulted in a major uptake of IPM and other sustainable agricultural practices. Some certifications, such as Fairtrade, in particular, also address soil issues thus helping farmers receive higher income (Kilian *et al.* 2006). However, quality of a product still has a greater bearing on the final price of a product than certification (Kilian *et al.* 2006). In the end, consumers demand healthy, high quality food (Sharma *et al.* 2006) so optimal farm management and technologies which encourage the production of high quality food will remain critical (Kilian *et al.* 2006).

The above technologies and certification systems will help move agriculture towards sustainability. Ecological taxes in the North are also necessary to place a true cost on agricultural products, particularly meat, so that consumers limit their consumption of meat in preference for fruit, vegetables and cereals. This will reduce the need to convert additional natural habitats to arable land to be used as animal feedstuffs and help deal with the obesity problem in the North.

The issue of trade liberalisation and North-South relations is perhaps much more contentious and harder to resolve, particularly equitable food distribution. Some, like Ho and Ulanowicz (2005), are not concerned with growing populations and even call for higher population densities, which they believe, are necessary to provide labour to manage the natural resources. Organisms are the most energy efficient 'machines' by far (Ho and Ulanowicz, 2005). Shah (2006) also agrees with this point of view and suggests that trade liberalisation should allow for labour movement from the South to North to improve the management of natural resources in the North (Shah, 2006). Such measures may alleviate some problems but also create new ones and are unlikely to receive the blessing of politicians in the North.

5 Conclusion

Agriculture should not simply "return to the past" but must find a way forward which includes learning from the past and local indigenous practices as well as by adopting green technologies that will encourage lower inputs, increased efficiency and increased yields. Increased yields are essential to meet the demands of the growing population. But just as importantly, these yields must be increased sustainably, so that environmental quality is maintained and improved over time. Sustainable agriculture must integrate the three main goals: economic, environmental and social; and must strive to meet the Millennium Development Goals, which seek to reduce malnutrition while maintaining environmental quality. This is not an easy task.

More research is needed in exploring sustainable development options in agriculture, particularly for the development of green technologies. Green technologies can be encouraged through certification schemes such as EUREPGAP, Faitrade and Organic, however greater awareness raising amongst consumers about sustainable agriculture is also necessary to encourage them to purchase such products. Enacting some of these standards in legislation in the North would greatly improve natural resource management both in the North and South. Ecological taxes that will increase the price of meat are also a necessary measure to reduce meat consumption and therefore the areas used for the growing of feedstuffs.

Some social issues can also be resolved through certification mechanisms; however, equitable food distribution would require a broader policy shift and agreement at the international level.

6 References

Birch, C. (1993). *Confronting the Future*. Penguin Books Australia: Ringwood, Victoria.

Borch, K. (2007). Emerging technologies in favour of sustainable agriculture. *Journal of Futures*. 2007; doi:10.1016/j.futures.2007.03.016

Bouwmann, A. (2001). Global estimates of gaseous emissions from agricultural land. FAO: Rome.

Bruinsma, J.E. (2003) World Agriculture: Towards 2015/2030. In FAO Perspective. Earthscan: London.

Dalvie, M.A., Africa, A., London, L. (2006). Disposal of unwanted pesticides in Stellenbosch, South Africa. *Science of the Total Environment*. 2006; 361, p8-17.

Ehrlich, A. (1994). Building a sustainable food system <u>in</u> Smith, P.B. *et al.* (eds) (1994) *The World at Crossroads.* A report to the Pugwash Council. Earthscan Publication: London.

Eickhout, B., Bouwman, A.F., van Zeijts, H. (2006). The role of nitrogen in world food production and environmental sustainability. *Agriculture, Ecosystems and Environment*. 2006; 111, p4-14.

Ho, Mae-Wan, Ulanowicz, R. (2005). Sustainable systems as organisms? *BioSystems*. 2005; 82, p39-51.

Hole, D.G., Perkins, A.J., Wilson, J.D., Alexander, I.H., Grice, P.V., Evans, A.D. (2005). Does organic farming benefit biodiversity? *Biological Conservation*. 2005; 122, p113-130.

Hoogland, C.T., de Boer, J., Boersema, J.J. (2007). Food and sustainability: Do consumsers recognise, understand and value on-package information on production standards? *Appetite*. 2007; 49, p47-57.

IPPC (2001). Climate Change 2001: Synthesis Report. Summary for Policy Makers. IPPC: Webley, UK.

Kilian, B., Jones, C., Pratt, L., Villalobos, A. (2006). Is sustainable agriculture a viable strategy to improve farm income in Central America? A case study on coffee. *Journal of Business Research*. 2006; 59, p322-330.

Mozumber, P., Berrens, R.P. (2007). Inorganic fertilizer use and biodiversity risk: An empirical investigation. *Ecological Economics*. 2007; 62, p538-543.

Mwakalila S. (2006). Socio-economic impacts of irrigated agriculture in Mbarali Distrcit of south-west Tanzania. *Physics and Chemistry of the Earth.* 2006; 31, p876-884.

Ruttan, V.W. (1994). Constraints on the design of sustainable systems of agricultural production. *Ecological Economics.* 1994; 10, p209-219.

Shah, A. (2006). Exploring sustainable production systems for agriculture: Implications for employment and investment under north-south trade scenario. *Ecological Economics.* 2006; 59, p237-241.

Sharma, T., Carmichael, J., Klinkerberg, B. (2006). Integrated modeling for exploring sustainable agriculture futures. *Futures.* 38; p93-113.

Shi, T. (2002). Ecological agriculture in China: bridging the gap between rhetoric and practice of sustainability. *Ecological Economics.* 2002; 42, p359-368.

Smith, P., Martino, D., Cao, Z., Gwary, D., Janzen, H., Kumar, P., McCarl, B., ogle, S., O'Mara, F., Rice, C., Scholes, B., Sirotenko, O., Howden, M., McAllister, T., Pan, G., Romanenkov, V., Schneider, U., Towprayoon, S. (2007). Policy and technological constraints to implementation of greenhouse gas mitigation options in agriculture. *Agriculture, Ecosystems and Environment.* 2007; 118, p6-28.

Soule, J.D. and Piper, J.K. (1992). *Farming in Nature's Image.* Island Press; Washington.

Tait, J., Morris, D. (2000). Sustainable development of agricultural systems: competing objectives and critical limits. *Futures.* 2000; 32, p247-260.

Toledo, A., Burlingame, B. (2006). Biodiversity and nutrition: A common path toward global food security and sustainable development. *Journal of Food Composition and Analysis.* 2006; 19, p477-483.

UCLA (1997). What is sustainable agriculture? University of California, Sustainable Agriculture Research and Education Program. Downloaded from: http://www.sarep.ucdavis.edu/concept.htm on July 18, 2007.

Wesseling, C., Corriols, M., Bravo, V. (2005). Acute pesticide poisoning and pesticide registration in Central America. *Toxicology and Applied Pharmacology.* 2005; 207, pS697-S705.

Wichelns, D., Oster, J.D. (2006). Sustainable irrigation is necessary and achievable, but direct costs and environmental impacts can be substantial. *Agricultural Water Management.* 2006; 86, p114-127.